# START EXPLORING

# Farms

## FACTS • ACTIVITIES • FUN

## Jacqueline Dineen

### Illustrated by Rod Holt

## CONTENTS

Headway • Hodder & Stoughton

# On the farm

There are farms all over the world. Some are very small, and others are enormous.

This is a mixed farm.

Sheep are kept for their meat and their wool.

Cows give milk. They are brought in for milking twice a day.

Pigs are kept for meat. They can live outdoors or in pens called pigsties.

The farmer grows some crops. One of these is wheat. Wheat is **cereal**. The grains of wheat are ground to make flour. The farmer also grows potatoes and other vegetables.

Most of our eggs come from hens. Ducks and geese are also kept for eggs.

# Growing food

Farmers grow cereals,
vegetables and fruit.

In the spring, when the weather
is warmer, the plants begin to
grow.

The farmer sows the wheat
seeds in February or March if
the ground is not frozen.

You can buy sprouting seeds which grow very quickly. Put a handful of alfalfa or mung bean seeds in a jam jar. Cover with a muslin cloth fixed with a rubber band. Twice a day, pour in cold water through the cloth, shake and then drain the water out. Stand the jar on its side in a warm room. Your seed sprouts will be ready in 5 to 7 days.

The summer is the hottest time of the year. The wheat begins to ripen in the sunshine. It turns from green to yellow.

When the wheat is ready, it is harvested.

In the autumn, the weather becomes colder again. As soon as the harvest is over, the farmer ploughs the field. If the weather is fine, he may sow a winter crop of wheat now.

5

# Cows

A cow produces milk for her calves to drink.

When the calf is a few days old, the farmer starts to feed it on other foods, so that he can milk the cow. First he gives the calf a milky mixture. When it is a bit older he feeds it with hay.

If a cow has a calf every year, she can give milk nearly all the time. The milk is stored in her **udder**. Milking machines suck the milk out.

## AMAZING FACTS!

A cow eats about 95 kilograms of grass and drinks 70 litres of water each day. The grass and water are turned to milk in her stomach. She gives about 22 litres of milk each day.

Milk soon turns sour if it is not pasteurised at the dairy. Pasteurisation was discovered by a French scientist called Louis Pasteur, who found that the germs which turn milk sour can be killed by heating the milk and then cooling it quickly.

The milking machine has four cups which fit on to the cow's teats. To the cow, they feel like a calf **suckling**.

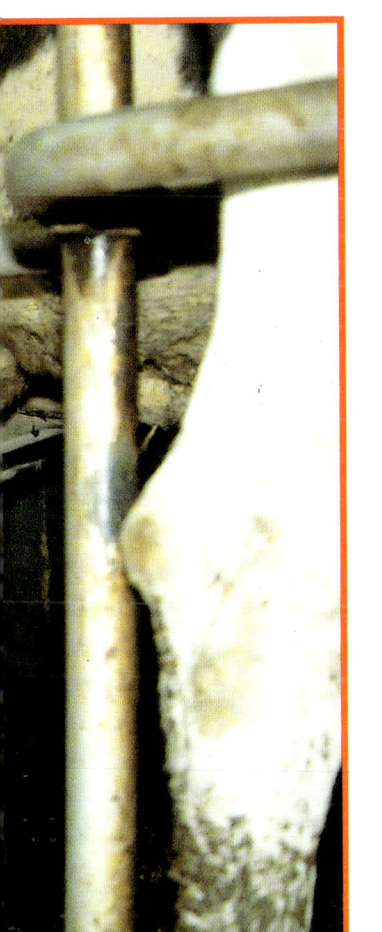

## Now You See

Butter and cheese are made from milk or cream. You can try making some butter. Pour the cream from two bottles of milk into a clean jam jar. Screw the lid on tightly.

Shake the jar hard. What happens? After a while, the fat sticks together in a lump. This is butter. Pour away any liquid and try your butter.

# Sheep

Sheep have soft curly wool which can be twisted or spun into long threads and made into cloth. They are also kept for meat.

There are 1,000 million sheep in the world. In many countries farmers keep a flock of sheep on a mixed farm. In Australia and New Zealand there are enormous sheep farms. The biggest ones may have 100,000 sheep. There are 68 million sheep in New Zealand - that is 20 times as many sheep as people!

Farmers **shear** the sheeps' wool in the spring when the weather is warm. This gives time for the wool to grow again before winter arrives. The shearer uses electric clippers to take off the whole coat, or **fleece**, in one piece.

# AMAZING FACTS!

A shearer can shear a sheep in 5 minutes or 125 sheep in a day. The hand-shearing record is held by a shearer in New Zealand who sheared 353 lambs in 9 hours. That is one lamb every 1.5 minutes!

# Eggs

All female birds lay eggs. If the bird has mated with a male, a baby bird develops inside each egg she lays. The bird sits on the eggs until the chicks hatch out.

You are not killing a chick when you eat an egg. Hens on egg farms are not mated with a male, so there are no chicks inside the eggs.

Hens do not lay eggs in winter when it is cold and dark. If a farmer wants to produce eggs all year round, the hens have to be kept in a deep-litter house. This is warm and has artificial light so the hens think spring has arrived.

Most of the eggs you buy come from battery farms. Thousands of hens are kept in huge sheds. They live in cages and cannot move about. They just lay eggs all day long.

You can buy 'free-range' eggs laid by hens which are left to peck around in the garden or farmyard.

# AMAZING FACTS!

The Ancient Egyptians ate ostrich eggs, and people in South Africa still eat them today. One ostrich egg weighs 1.5 kg. That is the same as 24 hens' eggs!

11

# Other farm animals

Not many farms have horses these days because they are not needed to pull machines. Some farmers keep horses for the family to ride.

The farmer has a sheepdog to help round up the sheep. The dog runs around behind the sheep and chases them into a flock.

We get pork, ham, bacon and sausages from pigs. On some pig farms the pigs are kept indoors all the time. They are fed on special food to fatten them up. Sometimes pigs are kept outside. They root around for food and sleep in a simple shelter.

# AMAZING FACTS!

Pigs often have a lot of piglets at once. The record is 34 in one litter!

Mice and rats feed on grain stored in barns, so the farmer has some cats to kill them. The cats stay out at night, hunting.

# Farm machines

Early farmers had very simple tools. They could only farm small pieces of land. Many people had to be farmers and grow their own food.

Today, farmers have machines to help them. Machines can do the work far more quickly than people. The farms can be much larger.

Early farming machines were pulled by horses. On modern farms, a tractor pulls the plough and the seed drill. The seed drill sows the seeds.

# Now You See

On modern farms, cows are milked by machines. They used to be milked by hand like this. How do you think seeds were sown before the seed drill was invented? Draw a picture of your idea.

# The combine harvester

The combine harvester is a machine which cuts the crop and separates the grain from the stalks.

The machine has a reel on the front. As this turns, it feeds the wheat on to the cutters.

The stalks will become straw.

The cut crop passes through the harvester into the drum. Here, the grain is separated from the stalks.

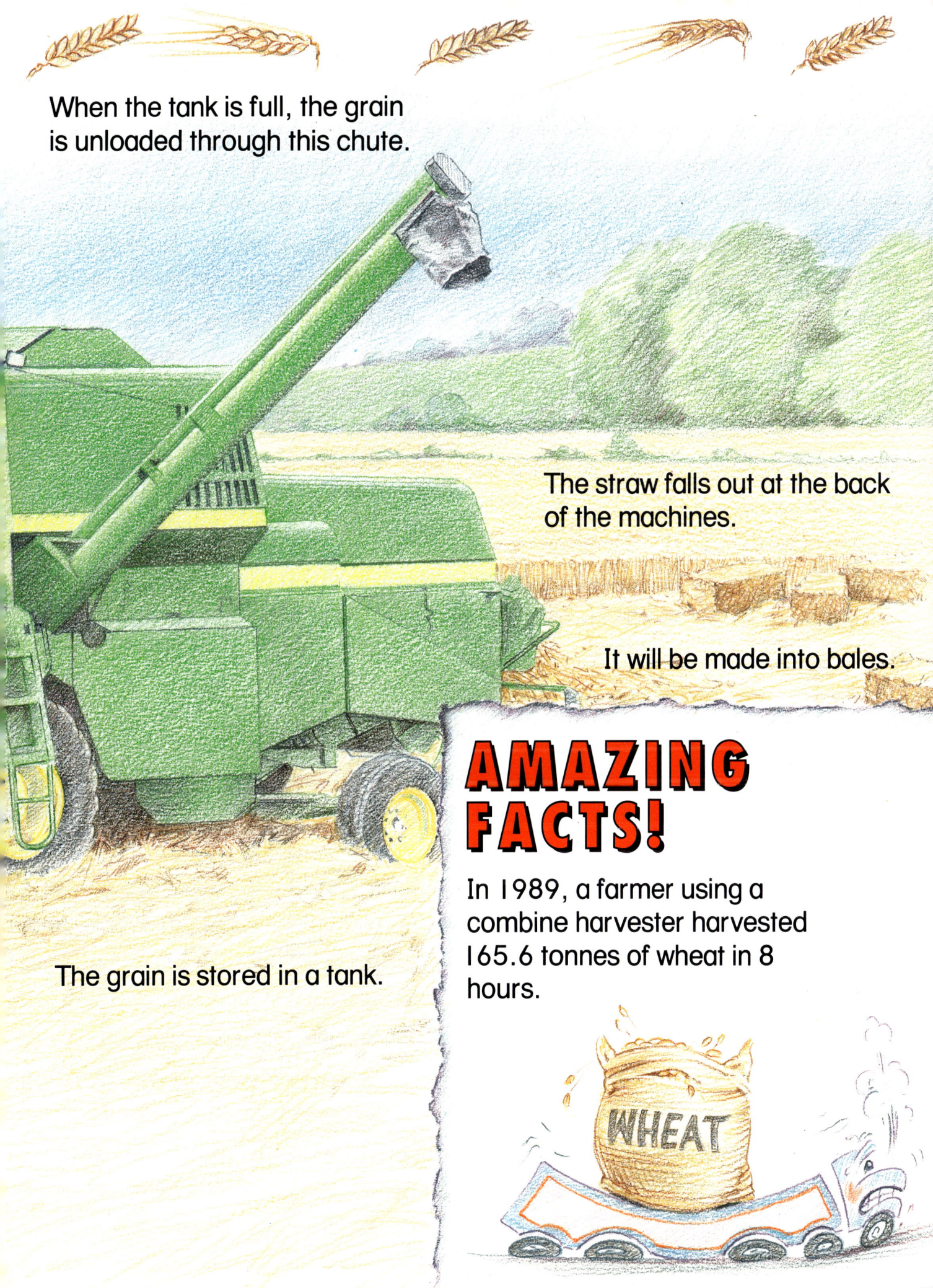

When the tank is full, the grain is unloaded through this chute.

The straw falls out at the back of the machines.

It will be made into bales.

The grain is stored in a tank.

## AMAZING FACTS!

In 1989, a farmer using a combine harvester harvested 165.6 tonnes of wheat in 8 hours.

WHEAT

# The farmer's day

The hens are pecking around in the yard. The farmer's daughter gives them some grain.

The farmer feeds the young calves. He gives the older calves hay so they have something to chew all day.

The hens have been shut up in a hen-house for the night. The farmer lets them out.

The farmer gets up at 6 o'clock. He has to bring the cows in for milking.
When he has finished the milking, he takes the cows back to the field.

18

The milk tanker arrives to collect the milk. The driver measures the amount before loading it into the tanker.

Today, the farmer is harrowing a field before sowing. Harrowing breaks the soil into smaller pieces after ploughing.

In the evening the cows are milked again.

## Now You See

What other things does the farmer have to do during the year? Draw a picture of one of your ideas.

# Wildlife on the farm

Because big machines cannot work in small fields, many hedges have been cut down to make the fields bigger.

The hedges were the **habitat** or home of small animals. Wild flowers grew in the grass. When the hedges were destroyed, the animals had nowhere to live. The flowers were pulled up.

Nowadays, some farmers are allowing the hedgerows to grow again.

Small animals such as mice, voles, hedgehogs and rabbits live in or near the hedge. It makes a safe home, away from animals which hunt them.

The hedge is made of shrubs such as hawthorn, elderberry and brambles. Honeysuckle twines its way between the shrubs.

Make your own hedgerow survey. What other habitats are there on a farm? What about barns and fields?

The birds which nest in the hedge include yellowhammers, blackbirds and chaffinches.

Wild flowers bloom in spring and summer. There are primroses, herb-Robert, meadowsweet, cornflowers, and many others.

# AMAZING FACTS!

In Brazil, an area of rain forest the size of a football pitch is cut down every second.

# Farms around the world

Farmers all over the world grow crops to sell to other countries. That is why you can buy fruit like bananas and pineapples in the shops even though they only grow in hot climates. Coffee, tea and cocoa are other crops which are grown in hot countries.

A cocoa farm in West Africa.

Picking coffee in South America.

Many people in Asia, Africa and South America farm small plots of land and can only grow enough food for their families. In some of these places the weather is very hot and dry for part of the year, then there is a long spell of rain. If the rains do not come, there is no water for the crops and they die.

Rice is grown in places which have a **monsoon** season, such as India and China. It rains non-stop from May to September. As soon as the rains begin, the farmer plants rice seedlings in the flooded fields. He harvests the rice when the fields have dried out after the rain.

A tea estate in India.

Growing bananas in Africa.

## Now You See

Which of these are fruit and which are vegetables?

They are all fruit because they all have seeds inside. Which ones grow in hot countries? Which ones grow in cooler countries?

# GLOSSARY

**cereal**   A grass plant which produces seeds that can be eaten.

**crops**   Plants which are grown by people for food.

**dairy**   A place where milk is treated to keep it fresh, and then put into bottles or cartons.

**fleece**   The complete wool coat of sheep, which is cut off in one piece.

**monsoon**   A strong wind that blows across south-east Asia and brings a rainy season.

**seedling**   A young plant.

**shear**   To cut off a sheep's wool.

**udder**   The bag-like part of animals such as cows, where milk is stored. The milk is sucked out through the teats.